Benefits of Following Manufacturer Maintenance Recommendations

Introduction

- **Importance of Car Maintenance**: Discuss the critical role of regular car maintenance in ensuring safety, performance, and longevity.
- **Overview of Manufacturer Recommendations**: Explain what manufacturer recommendations are and why they are essential.

Chapter 1: Understanding Manufacturer Recommendations

- **Definition and Purpose**: Define manufacturer maintenance recommendations and their purpose.
- **Development Process**: Describe how manufacturers determine these recommendations.
- **Normal vs. Severe Maintenance Schedules**: Explain the differences and why they matter.

Chapter 2: Enhanced Safety

- **Safety-Related Maintenance Tasks**: Detail tasks like brake inspections, tire checks, and light replacements.
- **Impact on Safety**: Discuss how these tasks prevent accidents and ensure safe driving.

Chapter 3: Improved Performance and Efficiency

- **Performance Benefits**: Explain how regular maintenance improves car performance.
- **Specific Maintenance Tasks**: Detail tasks such as oil changes, air filter replacements, and tire rotations.
- **Fuel Efficiency**: Discuss the impact of maintenance on fuel efficiency and cost savings.

Chapter 4: Cost Savings

- **Preventing Major Repairs**: Explain how regular maintenance can prevent costly repairs.
- **Cost Comparison**: Provide a comparison of costs between regular maintenance and major repairs.
- **Case Studies**: Include real-life examples of cost savings through regular maintenance.

Chapter 5: Longevity and Reliability

- **Extending Vehicle Lifespan**: Discuss how maintenance extends the life of a vehicle.
- **Engine Health**: Explain the importance of maintaining engine health.

- **High-Mileage Stories**: Share stories of cars that have achieved high mileage through regular maintenance.

Chapter 6: Resale Value

- **Impact on Resale Value**: Explain how maintenance records affect a car's resale value.
- **Keeping Records**: Provide tips for keeping detailed maintenance records.

Chapter 7: Environmental Benefits

- **Reduced Emissions**: Discuss how proper maintenance reduces emissions.
- **Sustainable Driving**: Explain the role of maintenance in promoting sustainable driving practices.
- **Environmental Impact**: Highlight the broader environmental benefits of maintaining a car.

Chapter 8: DIY Maintenance Tips

- **Basic Maintenance Tasks**: List tasks that car owners can do themselves.
- **Tools and Resources**: Provide information on the tools and resources needed for DIY maintenance.
- **Step-by-Step Guides**: Include step-by-step guides for common maintenance tasks.

Chapter 9: Choosing the Right Service Provider

- **Finding a Reliable Mechanic**: Offer tips on how to find a trustworthy mechanic.
- **Questions to Ask**: List important questions to ask a service provider.
- **Using OEM Parts**: Explain the importance of using Original Equipment Manufacturer (OEM) parts.

Conclusion

- **Recap of Benefits**: Summarize the benefits of following manufacturer recommendations.
- **Encouragement**: Encourage readers to stay on top of their car maintenance.
- **Final Thoughts**: Provide final thoughts and a call to action for readers to prioritize car maintenance.
- **Testimonials**: Include testimonials from car owners and dealers about the importance of maintenance records.

Importance of Car Maintenance:

Regular car maintenance is not just an option; it's a necessity. It plays a **pivotal role** in ensuring the **safety**, **performance**, and **longevity** of your vehicle. Here's why:
1. **Safety**: Regular maintenance checks help identify potential issues before they become serious problems. From brake inspections to tire pressure checks, these routine tasks ensure that your vehicle is safe to operate, protecting you, your passengers, and other road users.
2. **Performance**: Regular servicing keeps your car running at its optimal level. Tasks like oil changes, air filter replacements, and spark plug inspections contribute to better fuel efficiency, smoother rides, and improved overall performance.
3. **Longevity**: Regular maintenance extends the lifespan of your vehicle. By addressing minor issues early, you can prevent major, costly repairs down the line. This means your car stays on the road longer, providing better value for your investment.

In conclusion, regular car maintenance is an essential practice for every car owner. It's not just about keeping your car running; it's about ensuring a safe, efficient, and long-lasting vehicle. Remember, a well-maintained car is a well-performing car.

Overview of Manufacturer Recommendations:
Manufacturer recommendations are guidelines provided by the manufacturer of a product, particularly complex machinery like vehicles or appliances. These recommendations typically include:

1. **Maintenance schedules**: This includes when to perform certain checks or replacements, such as oil changes, tire rotations, or filter replacements in a car.
2. **Usage instructions**: This covers how to use the product correctly and efficiently to ensure optimal performance and longevity.
3. **Safety precautions**: These are measures to prevent accidents or damage while using the product.

Following manufacturer recommendations is essential for several reasons:

- **Product longevity**: Proper maintenance and usage as per the manufacturer's guidelines can significantly extend the life of the product.
- **Optimal performance**: The product will perform at its best when used as intended by the manufacturer.
- **Safety**: Adhering to the manufacturer's safety precautions helps prevent accidents and injuries.
- **Warranty validity**: Many manufacturers require customers to follow their recommendations to keep the product's warranty valid.

Chapter 1 Understanding Manufacturer Recommendations:

Manufacturer maintenance recommendations are specific guidelines provided by the manufacturer of a product, particularly complex machinery like vehicles or appliances. These guidelines outline the necessary steps to maintain the product's optimal performance and longevity.

For example, in the context of a car, these recommendations might include:

1. **Scheduled Servicing**: This includes regular checks and replacements such as oil changes, brake checks, tire rotations, and filter replacements. The manufacturer often provides a schedule based on mileage or time.
2. **Inspections**: Certain parts of the product may need regular inspections to ensure they are in good working condition. For instance, a car's belts, hoses, and fluids may need to be inspected at certain intervals.
3. **Cleaning and Care**: This involves instructions on how to clean and care for the product to prevent damage and wear.

The purpose of these recommendations is multi-fold:

- **Preserve Product Life**: Regular maintenance as per the manufacturer's guidelines can significantly extend the life of the product.
- **Ensure Optimal Performance**: Maintenance helps the product to perform at its best, providing the user with the expected level of service.
- **Prevent Breakdowns and Repairs**: Regular maintenance can help identify potential issues before they become serious problems, saving time and money on repairs.
- **Safety**: Regular maintenance can also ensure the product operates safely, preventing accidents or injuries.
- **Maintain Warranty**: In many cases, adherence to the manufacturer's maintenance recommendations is necessary to keep the product's warranty valid.

In summary, manufacturer maintenance recommendations are designed to help users maintain their products effectively, ensuring optimal performance, safety, and longevity. They are based on the manufacturer's extensive knowledge and testing of the product, making them a reliable guide for maintenance.

Manufacturers determine maintenance recommendations through a rigorous process that involves extensive research, testing, and analysis. Here's a general overview of how they do it:

1. **Product Testing**: Manufacturers conduct thorough testing of the product under various conditions to understand its performance and durability. This includes stress tests, endurance tests, and usage tests.
2. **Data Analysis**: The data collected from these tests is then analyzed to identify patterns and trends. For instance, they might find that a certain part tends to wear out after a specific number of uses or a certain period.
3. **Expert Review**: Engineers and experts review the data and use their expertise to determine the optimal maintenance schedule. They consider factors such as average usage patterns, environmental conditions, and safety considerations.
4. **Field Data**: Manufacturers also consider data from real-world usage of their products. This includes customer feedback, warranty claims, and service records. This data can provide valuable insights into how the product performs over time in various conditions.
5. **Continuous Improvement**: As manufacturers gather more data and feedback, they continuously update their maintenance recommendations to ensure they remain effective and relevant.

It's important to note that these recommendations are designed to ensure the product operates safely and efficiently over its expected lifespan. They are based on averages and typical usage patterns, so individual experiences may vary. Users should always consult with professionals for personalized advice.

Normal and Severe Maintenance Schedules are two types of maintenance plans recommended by manufacturers for their products, particularly vehicles. The choice between these two depends on the conditions in which the product, such as a car, is used.

Normal Maintenance Schedule: This schedule is designed for vehicles that operate under ideal or 'normal' conditions. These conditions may include:

- Moderate temperatures and climate
- Minimal towing or heavy loading
- Mainly highway driving at steady speeds
- Operating in clean, dust-free environments

Severe Maintenance Schedule: This schedule is for vehicles that operate under more demanding or 'severe' conditions. These conditions may include:

- Extreme temperatures (either hot or cold)
- Frequent short trips (less than 5 miles)
- Prolonged idling or stop-and-go traffic
- Heavy loads or towing
- Operating in dusty, muddy, or salty environments

The main difference between the two schedules lies in the frequency of maintenance tasks. Under a severe maintenance schedule, tasks like oil changes, filter replacements, and various inspections are performed more frequently due to the increased wear and tear.

Why do they matter?

- **Preserving Vehicle Health**: Following the appropriate maintenance schedule helps ensure the vehicle operates efficiently and lasts as long as possible.
- **Preventing Costly Repairs**: Regular maintenance can catch potential issues early before they turn into expensive repairs.
- **Safety**: Keeping up with maintenance helps ensure the vehicle operates safely.
- **Resale Value**: Vehicles with a well-documented maintenance history often have higher resale values.

It's important to consult your vehicle's owner's manual or a professional to determine which schedule is appropriate for your specific circumstances. Remember, what might be 'normal' conditions for one driver could be 'severe' for another. It's always better to err on the side of caution when it comes to vehicle maintenance.

Chapter 2: Enhanced Safety

1. **Brake Inspections**: Brakes are crucial for vehicle safety, so regular inspections are important. During a brake inspection, a mechanic will check the brake pads or shoes for wear, inspect the rotors, check brake fluid levels, and look for any leaks in the system. They'll also check the brake lines and hoses for any signs of cracking or damage. If any issues are found, they may recommend repairs or replacements to ensure the brakes function properly.
2. **Tire Checks**: Regular tire checks can help ensure your vehicle's performance and safety. These checks typically include:
    - **Tire Pressure**: Checking the tire pressure ensures that your tires are inflated to the manufacturer's recommended PSI (pounds per square inch). Proper inflation can improve fuel efficiency and handling and prevent premature tire wear.
    - **Tread Depth**: The tread on your tires should be checked regularly to ensure its within safe limits. If the tread is too worn, it can affect the vehicle's grip, especially in wet conditions.
    - **Tire Rotation**: Tires should be rotated regularly to ensure even wear. This can extend the life of your tires and improve vehicle handling.
3. **Light Replacements**: Vehicle lights, including headlights, brake lights, turn signals, and reverse lights, are vital for safety. Regular checks can ensure they're functioning correctly. If a bulb is out, it should be replaced promptly. It's also a good idea to clean the lights regularly, as dirt and grime can reduce their effectiveness.

Remember, always refer to your vehicle's owner's manual for specific maintenance recommendations or consult with a professional mechanic. Regular maintenance can help keep your vehicle running smoothly and safely. It can also prevent more costly repairs down the line.

Brake Inspections and Maintenance

- **Importance**: Brakes are crucial for stopping your vehicle safely. Regular inspections ensure they are functioning correctly.
- **Tasks**: Checking brake pads, rotors, and fluid levels.

- **Prevention**: Worn-out brakes can lead to longer stopping distances or brake failure, increasing the risk of accidents. Regular maintenance ensures your brakes respond effectively in emergencies.

Tire Checks and Rotations

- **Importance**: Tires are the only contact point between your car and the road. Properly maintained tires ensure good traction and handling.
- **Tasks**: Checking tire pressure, tread depth, and rotating tires.
- **Prevention**: Under-inflated or worn-out tires can cause blowouts or loss of control, especially in adverse weather conditions. Regular checks and rotations help maintain optimal tire performance and safety.

Light and Signal Inspections

- **Importance**: Lights and signals are essential for visibility and communication with other drivers.
- **Tasks**: Inspecting and replacing headlights, taillights, turn signals, and brake lights.
- **Prevention**: Faulty lights can reduce visibility at night or in poor weather, and non-functioning signals can lead to miscommunication with other drivers, increasing the risk of collisions.

Steering and Suspension Maintenance

- **Importance**: The steering and suspension systems are vital for vehicle control and comfort.
- **Tasks**: Inspecting and maintaining steering components, shocks, and struts.
- **Prevention**: Worn or damaged components can lead to poor handling, increased stopping distances, and loss of control, especially during sudden maneuvers or on rough roads.

Fluid Checks and Replacements

- **Importance**: Various fluids (engine oil, brake fluid, transmission fluid, coolant) are essential for the smooth operation of your vehicle.
- **Tasks**: Regularly checking and replacing fluids as needed.
- **Prevention**: Low or dirty fluids can cause engine overheating, brake failure, or transmission issues, leading to breakdowns or accidents.

Windshield Wiper and Washer Maintenance

- **Importance**: Clear visibility is crucial for safe driving, especially in adverse weather conditions.
- **Tasks**: Inspecting and replacing wiper blades, checking washer fluid levels.
- **Prevention**: Worn wiper blades or low washer fluid can impair visibility during rain or snow, increasing the risk of accidents.

Battery Maintenance

- **Importance**: A reliable battery ensures your car starts and all electrical systems function properly.
- **Tasks**: Checking battery charge, cleaning terminals, and replacing old batteries.
- **Prevention**: A dead battery can leave you stranded in unsafe locations or situations. Regular maintenance ensures your battery is always ready to go.

By staying on top of these safety-related maintenance tasks, you can significantly reduce the risk of accidents and ensure a safer driving experience for yourself and others on the road. Regular maintenance not only keeps your car in good condition but also provides peace of mind knowing that your vehicle is safe to drive.

Chapter 3: Improved Performance and Efficiency

Regular car maintenance significantly enhances vehicle performance in several ways:

1. **Improved Fuel Efficiency**: Regularly changing the oil, replacing air filters, and ensuring proper tire pressure can help your car run more efficiently. This means better gas mileage and lower fuel costs.
2. **Enhanced Safety**: Routine checks on brakes, tires, and other critical components ensure they are functioning correctly, reducing the risk of breakdown and accidents.
3. **Prolonged Vehicle Lifespan**: Consistent maintenance helps prevent wear and tear, ensuring your car runs smoothly for many years.
4. **Better Handling and Performance**: When your car's components are well-maintained, it results in more responsive handling and a smoother ride. This makes driving more enjoyable and ensures your vehicle performs at the best.
5. **Reduced Long-Term Repair Costs**: Regular maintenance helps identify issues early, preventing them from developing into major, more expensive repairs. This can save you significant amount of money in long run.

Oil Changes

- **Frequency**: Typically, every 5,000 to 10,000 miles, but check your owner's manual for your car's specific recommendation.
- **Purpose**: Oil lubricates the engine's moving parts, reducing friction and preventing overheating. Over time, oil breaks down and becomes less effective.
- **Process**: Drain the old oil, replace the oil filter, and refill with fresh oil. This helps maintain engine performance and longevity.

2. Air Filter Replacements

- **Frequency**: Every 12,000 to 15,000 miles, or as recommended by your vehicle's manufacturer.
- **Purpose**: The air filter prevents dirt, dust, and debris from entering the engine. A clean air filter ensures optimal air flow and improves fuel efficiency.
- **Process**: Remove the old air filter and replace it with a new one. This is usually a quick and easy task that can be done during an oil change.

3. Tire Rotations

- **Frequency**: Every 6,000 to 8,000 miles, or during every other oil change.
- **Purpose**: Rotating tires ensure even wear, which extends the life of the tires and improves handling and safety.
- **Process**: Move the tires from one position to another (e.g., front to back, side to side). This helps maintain balanced traction and prevents uneven tread wear.

Regularly performing these maintenance tasks can significantly enhance your car's performance and longevity. If you have any other specific tasks or need more detailed instructions, feel free to ask!

Regular maintenance has a significant impact on fuel efficiency and can lead to substantial cost savings over time. Here's how:

1. Oil Changes

- **Impact on Fuel Efficiency**: Fresh oil reduces friction in the engine, allowing it to run more smoothly and efficiently. This can improve your car's fuel economy.
- **Cost Savings**: Improved fuel efficiency means you spend less on gas. Regular oil changes also prevent engine wear, avoiding costly repairs.

2. Air Filter Replacements

- **Impact on Fuel Efficiency**: A clean air filter ensures that the engine gets the right amount of air, which is crucial for optimal combustion. This can enhance fuel efficiency by up to 10%.
- **Cost Savings**: Better fuel efficiency translates to fewer trips to the gas station. Additionally, a clean air filter helps maintain engine health, reducing the likelihood of expensive repairs.

3. Tire Rotations and Proper Inflation

- **Impact on Fuel Efficiency**: Properly inflated and rotated tires reduce rolling resistance, which means the engine doesn't have to work as hard. This can improve fuel efficiency by up to 3%.
- **Cost Savings**: Maintaining correct tire pressure and regular rotations extend tire life, saving you money on replacements. Plus, better fuel efficiency means lower fuel costs.

Spark Plug Replacements

- **Impact on Fuel Efficiency**: Worn-out spark plugs can cause misfires, reducing fuel efficiency. Replacing them ensures efficient combustion.
- **Cost Savings**: Efficient combustion means better fuel economy and fewer emissions, saving you money on fuel and potential fines for emissions violations.

5. Regular Engine Tune-Ups

- **Impact on Fuel Efficiency**: Tune-ups ensure that all engine components are working optimally. This can significantly improve fuel efficiency.
- **Cost Savings**: A well-tuned engine runs more efficiently, reducing fuel consumption and preventing costly breakdowns.

By keeping up with these maintenance tasks, you can ensure your car runs more efficiently, saving you money on fuel and reducing the likelihood of expensive repairs. Do you have any other questions about car maintenance or fuel efficiency?

Chapter 4: Cost Savings

Regular maintenance is crucial in preventing major, costly repairs. Here's how it helps:

1. Early Detection of Issues

- **Benefit**: Routine inspections and maintenance allow mechanics to identify and address minor issues before they escalate into major problems.
- **Example**: A small oil leak can be fixed easily and inexpensively if caught early. If ignored, it could lead to engine damage, requiring costly repairs or even engine replacement.

2. Preventing Wear and Tear

- **Benefit**: Regular maintenance tasks, such as oil changes and tire rotations, reduce wear and tear on critical components.
- **Example**: Regular oil changes keep the engine lubricated, preventing excessive wear on engine parts. This helps avoid expensive repairs related to engine failure.

Maintaining Optimal Performance

- **Benefit**: Keeping components like the air filter, spark plugs, and fuel system clean and in good condition ensures the engine runs efficiently.
- **Example**: Replacing a clogged air filter can prevent the engine from working harder than necessary, which can lead to overheating and potential engine damage.

4. Extending Component Lifespan

- **Benefit**: Regular maintenance extends the lifespan of various car components, delaying the need for replacements.
- **Example**: Properly inflated and rotated tires wear evenly, extending their life and preventing premature replacement costs.

5. Avoiding System Failures

- **Benefit**: Regular checks on systems like brakes, cooling, and transmission help prevent sudden failures that can be expensive to fix.
- **Example**: Regular brake inspections can catch worn brake pads early, preventing damage to the rotors and more costly brake system repairs.

6. Ensuring Fluid Levels and Quality

- **Benefit**: Regularly checking and replacing fluids (oil, coolant, brake fluid) ensures they are at optimal levels and quality, preventing system failures.

- **Example**: Maintaining proper coolant levels and quality prevents overheating, which can cause severe engine damage.

By staying on top of regular maintenance, you can avoid unexpected breakdowns and the high costs associated with major repairs. This proactive approach not only saves money but also ensures your car remains reliable and safe to drive. If you have any specific maintenance concerns or need advice on a particular issue, feel free to ask!

Cost comparison

Regular Maintenance Costs (Approx)
1. **Oil Changes**
   - **Cost**: $30-$70 per change
   - **Frequency**: Every 3,000 to 5,000 miles
   - **Annual Cost**: Approximately $120-$280
2. **Air Filter Replacements**
   - **Cost**: $20-$50 per filter
   - **Frequency**: Every 12,000 to 15,000 miles
   - **Annual Cost**: Approximately $20-$50
3. **Tire Rotations**
   - **Cost**: $20-$50 per rotation
   - **Frequency**: Every 6,000 to 8,000 miles
   - **Annual Cost**: Approximately $40-$100
4. **Brake Inspections and Pad Replacements**
   - **Cost**: $100-$300 for brake pad replacement
   - **Frequency**: Every 20,000 to 50,000 miles
   - **Annual Cost**: Approximately $100-$300 (depending on driving habits)

Major Repair Costs
1. **Engine Repair or Replacement**
   - **Cost**: $3,000-$7,000 or more
   - **Cause**: Often due to neglected oil changes or overheating
2. **Transmission Repair or Replacement**
   - **Cost**: $1,800-$3,500 or more
   - **Cause**: Often due to lack of regular fluid changes and inspections
3. **Brake System Overhaul**
   - **Cost**: $300-$800 or more
   - **Cause**: Often due to neglected brake pad replacements leading to rotor damage
4. **Tire Replacement**
   - **Cost**: $400-$1,000 for a full set
   - **Cause**: Often due to uneven wear from lack of rotations and improper inflation

Summary of Savings

- **Regular Maintenance**: Annual cost of approximately $280-$730
- **Major Repairs**: Can range from $1,800 to $7,000 or more for a single repair

By investing in regular maintenance, you can avoid the high costs associated with major repairs. This proactive approach not only saves money but also ensures your car remains reliable and safe to drive.

Holman Fleet Maintenance Case Study

- **Background**: A company with a large fleet of vans and light-duty trucks struggled with tracking maintenance efficiency and costs.
- **Solution**: Holman implemented a tailored scorecard to monitor maintenance compliance, preferred vendor usage, and fuel usage trends.
- **Results**:
    - **Oil and Filter Savings**: Company saved $19,000 (11%) on lube oil and filter
    - **Tire Savings:** By switching to preferred national accounts, they saved $13,000 (5%) equivalents to 86 free tires.
    - **Fuel Savings:** Fuel exceptions decreased by 65%, resulting in $36,000 in fuel savings.

2. Micro Main Preventive Maintenance Study

- **Background**: A study was conducted to measure the return on investment (ROI) of preventive maintenance.
- **Solution**: The research team established a baseline of maintenance costs and implemented a preventive maintenance program.

These case studies demonstrate how regular maintenance can lead to substantial cost savings by preventing major repairs, improving efficiency, and extending the lifespan of vehicle components.

Chapter 5: Longevity and Reliability

Extending the lifespan of your vehicle involves a combination of regular maintenance, mindful driving habits, and timely repairs. Here are some key strategies:

1. Regular Maintenance

- **Oil Changes**: Keep up with regular oil changes to ensure your engine runs smoothly and efficiently.
- **Fluid Checks**: Regularly check and top off fluids, including coolant, brake fluid, transmission fluid, and power steering fluid.
- **Filter Replacements**: Replace air, fuel, and cabin filters as recommended by your vehicle's manufacturer.
- **Tire Care**: Rotate tires regularly, maintain proper tire pressure, and check for alignment issues.

2. Mindful Driving Habits

- **Smooth Acceleration and Braking**: Avoid rapid acceleration and hard braking to reduce wear on the engine, brakes, and tires.
- **Avoid Overloading**: Don't exceed your vehicle's weight capacity, as this can strain the engine, suspension, and brakes.
- **Warm Up the Engine**: Allow your engine to warm up for a few minutes before driving, especially in cold weather, to reduce wear.

Timely Repairs

- **Address Issues Promptly**: Fix minor issues as soon as they arise to prevent them from becoming major problems.
- **Use Quality Parts**: When replacing parts, use high-quality or OEM (Original Equipment Manufacturer) parts to ensure compatibility and longevity.
- **Regular Inspections**: Have your vehicle inspected by a professional mechanic regularly to catch potential issues early.

4. Protective Measures

- **Keep It Clean**: Regularly wash and wax your car to protect the paint and prevent rust.
- **Garage Parking**: Park in a garage or use a car cover to protect your vehicle from the elements.
- **Rust Prevention**: Apply rust-proofing treatments, especially if you live in an area with harsh winters or near the ocean.

Follow Manufacturer's Recommendations

- **Service Schedule**: Adhere to the maintenance schedule outlined in your vehicle's owner's manual.

- **Recalls and Updates**: Stay informed about any recalls or software updates for your vehicle and address them promptly.

By following these strategies, you can significantly extend the lifespan of your vehicle, ensuring it remains reliable and efficient for many years.

Optimal Performance

- **Smooth Operation**: A well-maintained engine runs smoothly, providing better acceleration, power, and fuel efficiency.
- **Consistent Power**: Regular maintenance ensures that the engine delivers consistent power, making your driving experience more enjoyable.

2. Fuel Efficiency

- **Reduced Fuel Consumption**: A healthy engine operates more efficiently, which means it uses less fuel. This can lead to significant cost savings over time.
- **Lower Emissions**: Efficient engines produce fewer emissions, contributing to a cleaner environment.

3. Preventing Major Repairs

- **Early Detection**: Regular maintenance helps identify potential issues before they become major problems, saving you from costly repairs.
- **Avoiding Breakdowns**: Keeping the engine in good condition reduces the risk of unexpected breakdowns, ensuring your vehicle remains reliable.

4. Prolonged Engine Life

- **Reduced Wear and Tear**: Regular oil changes, filter replacements, and other maintenance tasks reduce wear and tear on engine components, extending the engine's lifespan.
- **Preventing Overheating**: Proper maintenance ensures that the cooling system functions correctly, preventing the engine from overheating and sustaining damage.

Safety

- **Reliable Performance**: A well-maintained engine is less likely to fail, ensuring your vehicle remains safe to drive.
- **Avoiding Accidents**: Engine issues can lead to sudden loss of power or control, increasing the risk of accidents. Regular maintenance helps prevent such scenarios.

6. Resale Value

- **Higher Resale Value**: A vehicle with a well-maintained engine is more attractive to potential buyers, commanding a higher resale value.

- **Proof of Care**: Maintenance records demonstrate that the vehicle has been well cared for, instilling confidence in buyers.

Key Maintenance Tasks for Engine Health

- **Oil Changes**: Regularly change the oil and oil filter to keep the engine lubricated and running smoothly.
- **Air Filter Replacements**: Ensure the engine gets clean air for optimal combustion.
- **Spark Plug Replacements**: Replace spark plugs to maintain efficient combustion and prevent misfires.
- **Coolant Checks**: Maintain proper coolant levels to prevent overheating.
- **Timing Belt/Chain Inspections**: Replace the timing belt or chain as recommended to avoid engine damage.

By prioritizing engine health through regular maintenance, you can ensure your vehicle remains reliable, efficient, and safe to drive.

The Million-Mile Honda Accord

- **Owner**: Joe LoCicero
- **Vehicle**: 1990 Honda Accord
- **Mileage**: Over 1 million miles
- **Maintenance Routine**: Joe followed the maintenance schedule religiously, performing regular oil changes, replacing filters, and addressing any issues promptly. He also kept detailed records of all maintenance and repairs.
- **Outcome**: Joe's dedication to maintenance allowed his Honda Accord to reach over 1 million miles, a testament to the durability of the vehicle and the importance of regular upkeep.

The 500,000-Mile Toyota Tacoma

- **Owner**: Victor Sheppard
- **Vehicle**: 2007 Toyota Tacoma
- **Mileage**: Over 500,000 miles
- **Maintenance Routine**: Victor adhered strictly to the manufacturer's maintenance schedule, ensuring timely oil changes, tire rotations, and other necessary services. He also used high-quality parts and fluids.
- **Outcome**: Victor adhered strictly to the manufacturer's maintenance schedule, ensuring timely oil changes, tire rotations, and other necessary services, He also used high quality parts.

The 300,000-Mile Volvo P1800

- **Owner**: Irv Gordon
- **Vehicle**: 1966 Volvo P1800
- **Mileage**: Over 3 million miles
- **Maintenance Routine**: Irv was meticulous about maintenance, performing regular oil changes, replacing parts as needed, and keeping the car in top condition. He also drove the car regularly, which helped keep it in good working order.
- **Outcome**: Irv's Volvo P1800 holds the Guinness World Record for highest mileage on a non-commercial vehicle, demonstrating the incredible longevity that can be achieved with proper maintenance.

These stories highlight the significant impact that regular maintenance can have on a vehicle's lifespan. By following the manufacturer's maintenance schedule and addressing issues promptly, you can keep your car running smoothly for many years and miles.

Chapter 6: Resale Value

Maintaining detailed service records can significantly impact your car's resale value. Here's how:

1. Proof of Regular Maintenance

- **Benefit**: Service records provide tangible proof that the vehicle has been well-maintained according to the manufacturer's recommendations.
- **Impact**: Buyers are more confident in purchasing a car with documented maintenance, as it indicates the vehicle is less likely to have hidden issues.

2. Increased Buyer Confidence

- **Benefit**: Detailed records show that the car has been cared for responsibly, which can increase buyer trust.
- **Impact:** This often translates to a higher resale value, as buyers are willing to pay more for well-documented vehicle.

3. Higher Resale Value

- **Benefit**: Cars with complete service histories can command higher prices.
- **Impact:** A well-maintained vehicle with a documented service history can increase resale value by 10% to 20%.

4. Easier Negotiations

- **Benefit**: Having a complete service history can make negotiations smoother.
- **Impact:** Sellers can justify their asking price more effectively, reducing the likelihood of price haggling.

Warranty and Insurance Claims

- **Benefit**: Service records can help with warranty claims and insurance settlements.
- **Impact**: They provide evidence that the vehicle was properly maintained, which can be crucial for warranty coverage and accurate insurance claims.

6. Attracting Serious Buyers

- **Benefit**: Serious buyers often look for well-maintained vehicles with a full-service history.
- **Impact**: This can lead to quicker sales and potentially multiple offers, driving up the final sale price.

By keeping detailed maintenance records, you not only ensure your car runs smoothly but also maximize its resale value when the time comes to sell.

keeping detailed maintenance records is a great way to ensure your vehicle remains in top condition and retains its value. Here are some tips to help you maintain thorough and organized records:

1. Use a Dedicated Notebook or Binder

- **Benefit**: Keeps all records in one place, making them easy to find and review.
- **Tip**: Use dividers to separate different types of maintenance (e.g., oil changes, tire rotations, brake services).

2. Digital Records

- **Benefit**: Digital records are easy to store, search, and share.
- **Tip**: Use apps or software designed for vehicle maintenance tracking, such as My Carfax, AutoSys, or simply a spreadsheet.

3. Keep All Receipts and Invoices

- **Benefit**: Provides proof of services performed and parts purchased.
- **Tip**: Attach or scan receipts and invoices to your physical or digital records.

Record Details of Each Service

- **Benefit**: Detailed records help track what is done and when.
- **Tip**: Include the date, mileage, type of service, parts used, and the name of the service provider.

5. Log DIY Maintenance

- **Benefit**: Even if you perform maintenance yourself, it is important to document it.
- **Tip**: Note the date, mileage, and details of the work you did, along with any parts or fluids used.

6. Schedule Reminders

- **Benefit**: Helps you stay on top of regular maintenance tasks.
- **Tip**: Use calendar reminders or maintenance tracking apps to alert you when services are due.

7. Organize by Date or Mileage

- **Benefit**: Makes it easy to see the history and plan future maintenance.
- **Tip**: Sort records chronologically or by mileage intervals.

8. Include Manufacturer Recalls and Updates

- **Benefit**: Ensures your vehicle is up to date with safety and performance improvements.
- **Tip**: Keep a section for any recall notices or software updates performed on your vehicle.

Maintain a Summary Sheet

- **Benefit**: Provides a quick overview of all maintenance performed.
- **Tip**: Create a summary sheet listing major services and their dates/mileage for easy reference.

10. Backup Your Records

- **Benefit**: Protects your records from loss or damage.
- **Tip**: Keep digital copies of physical records and backup digital records to cloud storage or an external drive.

By following these tips, you can ensure your maintenance records are comprehensive, organized, and easily accessible. This will not only help you keep your vehicle in excellent condition but also enhance its resale value.

Here are some testimonials from car owners who have experienced the benefits of maintaining detailed service records when selling their vehicles:

1. John's Honda Civic

- **Testimonial**: "I kept meticulous records of every oil change, tire rotation, and repair for my 2010 Honda Civic. When it came time to sell, I was able to show potential buyers a complete history of the car's maintenance. This transparency gave them confidence in the car's condition, and I ended up selling it for $1,500 more than similar models without detailed records."
- **Impact**: John's experience highlights how detailed maintenance records can boost buyer confidence and increase resale value.

2. Sarah's Toyota Camry

- **Testimonial**: "I always made sure to follow the manufacturer's maintenance schedule for my Toyota Camry and kept all the receipts and service logs. When I decided to upgrade, the dealership offered me a higher trade-in value because they could see the car was well-maintained. It made the entire process smoother and more profitable."
- **Impact**: Sarah's story shows how maintaining records can lead to higher trade-in offers from dealerships.

3. Mike's Ford F-150

- **Testimonial**: "I used a maintenance tracking app to log all the services for my Ford F-150. When I listed it for sale, I included a printout of the maintenance history. Buyers

appreciated the detailed records, and I received multiple offers above my asking price. It was clear that the records made a significant difference."
- **Impact**: Mike's use of a maintenance tracking app demonstrates how digital records can enhance the resale process and attract serious buyers.

4. Lisa's Subaru Outback

- **Testimonial**: "I kept a binder with all the service records for my Subaru Outback. When I sold it, the buyer was impressed with the thorough documentation and felt reassured about the car's reliability. I was able to sell it quickly and for a great price."
- **Impact**: Lisa's organized approach to record-keeping helped her sell her car quickly and at a higher price.

These testimonials illustrate the significant impact that detailed maintenance records can have on a vehicle's resale value. By keeping thorough and organized records, you can increase buyer confidence, justify a higher asking price, and ensure a smoother selling process.

Chapter 7: Environmental Benefits

Proper maintenance of your vehicle not only enhances its performance and longevity but also has significant environmental benefits by reducing emissions. Here's how:

1. Lower Carbon Footprint

- **Impact**: Regular maintenance ensures that your vehicle runs efficiently, reducing the amount of fuel it consumes and, consequently, the carbon dioxide (CO_2) it emits.
- **Benefit**: Lower CO_2 emissions contribute to mitigating climate change and reducing your overhaul carbon footprint.

2. Improved Air Quality

- **Impact**: Well-maintained engines burn fuel more completely, reducing the emission of harmful pollutants such as carbon monoxide (CO), nitrogen oxides (NO_x), and hydrocarbons (HC).
- **Benefit**: Reduced emissions of the pollutants improve air quality, which is beneficial for public health and the environment.

3. Reduced Fuel Consumption

- **Impact**: Efficiently running vehicles consume less fuel, which means fewer emissions are produced per mile driven.
- **Benefit:** Lower fuel consumption not only saves money but also reduces the demand for fossil fuels, leading to fewer emissions from fuel production and transportation.

Extended Vehicle Life

- **Impact**: Regular maintenance extends the lifespan of your vehicle, delaying the need for manufacturing new vehicles and the associated environmental impact.
- **Benefit**: This reduces the environmental footprint associated with the production, transportation, and disposal of vehicles.

5. Compliance with Emission Standards

- **Impact**: Keeping your vehicle in good condition ensures it meets emission standards set by regulatory bodies.
- **Benefit**: Compliance with these standards help reduce the overall environmental impact of transportation and supports cleaner air initiatives.

6. Waste Reduction

- **Impact**: Proper maintenance reduces the likelihood of major breakdowns and the need for extensive repairs, which can generate significant waste.
- **Benefit:** Minimizing waste from the repairs and replacements contributes to better waste management and less environmental pollution.

By maintaining your vehicle properly, you not only ensure its optimal performance but also contribute to a healthier environment. This proactive approach helps reduce harmful emissions, conserve resources, and promote sustainability.
Fuel efficiency has a significant impact on the environment. Here are some key points to consider:

1. **Reduced Greenhouse Gas Emissions**

   - **Impact**: Fuel-efficient vehicles emit fewer greenhouse gases (GHGs) such as carbon dioxide (CO_2), which are major contributors to climate change.
   - **Benefit**: Lower GHG emissions help mitigate global warming and its associated effects, such as extreme weather events and rising sea levels.

2. **Conservation of Non-Renewable Resources**

   - **Impact**: Fuel-efficient vehicles consume less fuel, reducing the demand for non-renewable resources like oil.
   - **Benefit**: This conservation helps preserve these resources for future generations and reduces the environmental impact of extraction and refining process.

Improved Air Quality

   - **Impact**: Efficient fuel combustion in fuel-efficient vehicles results in fewer pollutants like nitrogen oxides (NOx) and particulate matter (PM).
   - **Benefit**: Improved air quality leads to better public health outcomes, reducing respiratory and cardiovascular diseases.

4. **Lower Fuel Consumption**

   - **Impact**: Vehicles that use less fuel per mile driven reduce overall fuel consumption.
   - **Benefit**: This not only saves money for the consumers but also decreases the environmental footprint associated with fuel production and transportation.

5. **Reduced Dependence on Fossil Fuels**

   - **Impact**: Increased fuel efficiency reduces the reliance on fossil fuels, promoting energy independence.
   - **Benefit**: The shift supports the transition to cleaner energy sources and reduces the geopolitical and environmental risks associated with the fossil fuel dependence.

Eco-Driving Practices

   - **Impact**: Adopting eco-driving practices, such as smooth acceleration and braking, can further enhance fuel efficiency.
   - **Benefit**: These practices can reduce fuel consumption by 15% to 25% and GHG emissions by at least 30%.

By prioritizing fuel efficiency, we can make a significant positive impact on the environment, promoting sustainability and improving public health.

Proper maintenance plays a crucial role in sustainable driving by enhancing fuel efficiency, reducing emissions, and extending the lifespan of your vehicle. Here's how:

1. **Enhanced Fuel Efficiency**

   - **Impact**: Regular maintenance ensures that your vehicle operates at peak efficiency, using less fuel for the same distance.
   - **Benefit**: Improved fuel efficiency reduces the amount of fuel consumed, leading to lower greenhouse gas emissions and cost savings.

2. **Reduced Emissions**

   - **Impact**: Well-maintained vehicles emit fewer pollutants, such as carbon monoxide (CO), nitrogen oxides (NOx), and hydrocarbons (HC).
   - **Benefit**: Lower emissions contribute to better air quality and a healthier environment.

3. **Extended Vehicle Lifespan**

   - **Impact**: Regular maintenance prevents wear and tear, ensuring that your vehicle remains in good condition for a longer period.
   - **Benefit:** Extending the lifespan of your vehicle reduces the need for new car production, which has a significant environmental footprint.

4. **Proper Disposal of Automotive Fluids**

   - **Impact**: Responsible disposal of used oil, coolant, and other fluids prevents environmental contamination.
   - **Benefit**: Proper disposal practices protect water sources and wildlife from harmful chemicals.

5. **Eco-Friendly Maintenance Practices**

   - **Impact**: Using eco-friendly products and methods, such as biodegradable cleaning agents and water-saving car washes, minimizes environmental impact.
   - **Benefit:** These practices reduce pollution and conserve natural resources.

Compliance with Emission Standards

   - **Impact**: Regular maintenance ensures that your vehicle meets emission standards set by regulatory bodies.
   - **Benefit**: Compliance with these standards helps reduce the overall environmental impact of transportation.

By prioritizing regular maintenance, you can drive more sustainably, contributing to a cleaner environment and promoting long-term vehicle health.

Chapter 8: DIY Maintenance Tips

Here's a short list of basic car maintenance tasks you can do yourself:

1. **Oil Changes**: Replace the oil and oil filter.
2. **Air Filter Replacements**: Swap the old air filter for a new one.
3. **Fluid Checks**: Check and top off engine oil, coolant, brake fluid, and windshield washer fluid.
4. **Tire Maintenance**: Rotate tires and check tire pressure.
5. **Spark Plug Replacements**: Replace old spark plugs.
6. **Windshield Wiper Blades**: Install new wiper blades.
7. **Battery Maintenance**: Clean terminals and check battery charge.
8. **Interior Cleaning**: Vacuum and wipe down surfaces.

Here's a short list of essential tools and resources for DIY car maintenance:

1. **Basic Tool Set**: Includes wrenches, screwdrivers, pliers, and sockets.
2. **Jack and Jack Stands**: For safely lifting and supporting your car.
3. **Oil Filter Wrench**: To easily remove and replace oil filters.
4. **Tire Pressure Gauge**: To check and maintain proper tire pressure.
5. **Spark Plug Wrench**: For removing and installing spark plugs.
6. **Funnel**: For adding fluids without spilling.
7. **Multimeter**: To check battery voltage and electrical issues.
8. **Work Gloves**: To protect your hands while working.
9. **Service Manual**: Specific to your vehicle, providing detailed instructions.

These tools and resources will help you perform basic maintenance tasks effectively. Here are step-by-step guides for some common DIY car maintenance tasks:

1. Changing the Oil

1. **Warm Up the Engine**: Run the engine for a few minutes to warm the oil.
2. **Lift the Car**: Use a jack and jack stands or ramps to lift the car.
3. **Drain the Oil**: Locate the oil drain plug, place a drain pan underneath, and remove the plug to drain the oil.
4. **Replace the Oil Filter**: Remove the old oil filter and install a new one.
5. **Refill with Fresh Oil**: Replace the drain plug, lower the car, and refill with the recommended oil.

2. Replacing the Air Filter

1. **Locate the Air Filter Box**: Open the hood and find the air filter box.
2. **Remove the Old Filter**: Open the box and take out the old filter.
3. **Install the New Filter**: Place the new filter in the box and close it.

Checking and Topping Off Fluids

1. **Check Engine Oil**: Use the dipstick to check the oil level and add oil if needed.
2. **Check Coolant**: Look at the coolant reservoir and add coolant if the level is low.
3. **Check Brake Fluid**: Check the brake fluid reservoir and top off if necessary.
4. **Check Windshield Washer Fluid**: Fill the washer fluid reservoir if it's low.

4. Rotating Tires

1. **Lift the Car**: Use a jack and jack stands to lift the car.
2. **Remove the Tires**: Take off the tires and rotate them according to the recommended pattern (e.g., front to back, side to side).
3. **Reinstall the Tires**: Put the tires back on and tighten the lug nuts.

5. Replacing Spark Plugs

1. **Locate the Spark Plugs**: Open the hood and find the spark plugs.
2. **Remove the Old Spark Plugs**: Use a spark plug wrench to remove the old plugs.
3. **Install New Spark Plugs**: Insert the new plugs and tighten them.

Replacing Windshield Wiper Blades

1. **Lift the Wiper Arms**: Lift the wiper arms away from the windshield.
2. **Remove the Old Blades**: Press the release tab and slide the old blades off.
3. **Install New Blades**: Slide the new blades on until they click into place.

These guides can help you perform basic maintenance tasks and keep your car in good condition.

Chapter 9: Choosing the Right Service Provider

Finding a reliable mechanic is crucial for maintaining your vehicle's health. Here are some steps to help you find a trustworthy professional:

1. Ask for Recommendations

- **Benefit**: Personal referrals from friends, family, or colleagues can provide trustworthy insights.
- **Tip**: Ask about their experiences, the quality of work, and mechanic's reliability.

2. Check Online Reviews

- **Benefit**: Online reviews on platforms like Yelp, Google, and Car Talk's Mechanics Files can give you a broader perspective.
- **Tip: Look for mechanics with consistently high ratings and read through both positive and negative reviews to get balanced view.**

3. Verify Certifications

- **Benefit**: Certifications like ASE (Automotive Service Excellence) indicate that the mechanic has undergone rigorous testing and training.
- **Tip**: Look for certifications displayed in the shop or ask the mechanic directly about their qualifications.

4. Visit the Shop

- **Benefit**: A visit can give you a sense of the shop's professionalism and cleanliness.
- **Tip**: Observe the shop's organization, the condition of the equipment, and how the staff interacts with customers.

5. Ask About Warranties

- **Benefit**: A good mechanic will stand by their work and offer warranties on parts and labor.
- **Tip**: Inquire about the warranty policy to ensure you're covered in case of any issues.

6. Start with a Small Job

- **Benefit**: Testing a mechanic with a minor repair or maintenance task can help you gauge their competence and customer service.
- **Tip**: Use this opportunity to evaluate their communication, transparency, and the quality of their work.

7. Check for Memberships

- **Benefit**: Memberships in professional organizations, like AAA's Approved Auto Repair network, can be a good indicator of reliability.
- **Tip:** Look for shops that are part of a reputable network or association.

8. Get Multiple Estimates

- **Benefit**: Comparing estimates from different mechanics can help you understand the fair price for the work needed.
- **Tip:** Be wary of estimates that are significantly lower than others, as they might indicate subpar work or hidden cost.

By following these steps, you can find a reliable mechanic who will help keep your vehicle in excellent condition
Here's a list of important questions to ask your car service provider to ensure you receive quality service:

1. Have you worked on my make and model before?

- **Why Ask**: Ensures the mechanic has experience with your specific vehicle type.

2. What certifications do you have?

- **Why Ask:** Certifications like ASE (Automotive Service Excellence) indicates the mechanics expertise and training.

3. Can I get a written estimate?

- **Why Ask:** A written estimate helps you understand the costs involved and prevents unexpected charges.

4. What are my repair options?

- **Why Ask**: Knowing all available options allows you to make an informed decision about the repairs.

5. Are the parts or services covered by a warranty?

- **Why Ask:** Ensures you are protected if the repair doesn't resolve the issue or if part fails prematurely.

6. Can I see the old parts?

- **Why Ask:** Seeing old parts helps verify that replacements were necessary and builds trust.

What is your labor rate?

- **Why Ask**: Understanding the labor rate helps you compare costs with the other service providers.

8. Do you provide a detailed invoice?

- **Why Ask:** A detailed invoice breaks down the cost and services performed, ensuring transparency.

9. How do you handle unexpected issues?

- **Why Ask:** Knowing their process for dealing with unforeseen problems can help you avoid surprises.

10. Can I speak to other customers for references?

- **Why Ask**: References from other customers can provide insights into the quality of service and customer satisfaction.

These questions can help you find a reliable mechanic and ensure your vehicle receives the best possible care.

Here are the benefits of using OEM (Original Equipment Manufacturer) parts for your car:

1. **Exact Fit and Compatibility**
   - **Benefit**: OEM parts are designed specifically for your vehicle, ensuring a perfect fit and seamless integration with existing systems.
2. **Reliable Quality**
   - **Benefit**: These parts meet the exact specifications and standards by the vehicle manufacturer, guaranteeing consistent quality and performance.
3. **Warranty Assurance**
   - **Benefit**: OEM parts typically come with a warranty from the manufacturer, providing peace of mind and protection against defects.
4. **Performance Consistency**
   - Benefit: Using OEM parts ensures that the vehicle maintains its original performance levels, as these parts are designed to work harmoniously with your car.
5. **Safety**
   - **Benefit**: OEM parts are rigorously tested to meet the safety standards, ensuring that your vehicle remains safe to drive.
6. **Longevity and Durability**
   - Benefit: OEM parts are made from high quality materials, which often means they last longer and require less frequent replacements.
7. **Preserving Vehicle Value**
   - **Benefit**: Maintaining your vehicle with the OEM parts helps preserve its resale value, as buyers often prefer cars with original parts due to their known quality and compatibility.

8. **Avoiding Compatibility Issues**
    - **Benefit:** OEM parts eliminate the risk of compatibility issues that can arise with the aftermarket parts, ensuring smooth operation and reducing the likelihood of additional repairs.

By choosing OEM parts, you ensure that your vehicle remains reliable, safe, and maintains its value over time

Conclusion

In conclusion, adhering to manufacturer maintenance recommendations is a crucial step on the road to vehicle longevity. By following these guidelines, you ensure that your car operates at peak performance, maintains its safety standards, and retains its value over time. Regular maintenance tasks, such as oil changes, fluid checks, and tire rotations, not only enhance fuel efficiency and reduce emissions but also prevent costly repairs and extend the lifespan of your vehicle. Embracing these practices not only benefits your wallet but also contributes to a more sustainable and environmentally friendly driving experience. Ultimately, the commitment to regular maintenance is an investment in the reliability, safety, and longevity of your vehicle, ensuring many more miles of smooth and trouble-free driving.

www.ingramcontent.com/pod-product-compliance
Lightning Source LLC
Chambersburg PA
CBHW072056230526
45479CB00010B/1107